OFF GRID MOBILE SOLAR POWER INSTALLATION IN 1 HOUR

A Step by step Guide to Design and install 12 Volts Solar Power System on Vans, RVS, Boats and Mobile Homes

LARRY BARONE

Copyright

Larry Barone
ISBN: 9798657740226
ChurchGate Publishing House
USA | UK | Canada
© Churchgate Publishing House 2020

All rights reserved. No part of this publication may be reproduced, stored in a retrieval system or transmitted in any form or by any means, electronic, mechanical, photocopying, recording, and scanning without perm-ission in writing by the author.

While the advice and information in this book are believed to be true and accurate at the date of publication, neither the authors nor the editors nor the publisher can accept any legal responsibility for any errors or omissions that may be made. The publisher makes no warranty, express or implied, with respect to the material contained herein.

Printed on acid-free paper.

Printed in the United States of America
© 2020 by Larry Barone

Contents

Copyright i
Introduction to Electricity 1
What makes up electricity 2
The Forms of electricity 3
Measuring electricity 4
Electric Circuits: Serial vs Parallel 5
The difference between the parallel or serial circuit 6
The multiple advantages of the parallel circuit 7
Overview of Major Solar Panel System 8
Photovoltaic Solar System 9
Components of a solar power system 10
The regulator 11
The battery 11
The converter or inverter (optional) 12
How Does the System work? 12
Solar Panel System Design Methods 13
The minimalist 13
The classic 400 watt 14
The off Grid king 14
Ultra-lightweight 14
Low budget 15
Dystopian future 15
Traditional Method 15
Calculating the load 16

Calculating the battery bank size 16

Calculating Solar Array Size 17

How to calculate the maximum/ minimum solar array size for a battery 18

Other solar array sizing tips 18

Calculating Solar Charge Controller size 19

Efficiency Considerations 19

Other Factors To Consider 20

How To Select Solar Power System Components 21

Selecting a Battery 21

The main Technical Characteristics of Batteries For Solar Power Systems 23

Things To consider 23

Selecting Solar Panels 24

Installation of the flexible solar panel 26

Selecting a Solar Charge Controller 27

Selecting an inverter 29

The different models of solar inverter 30

Selecting wire 31

12 Volt Wire Gauge Chart 32

Between the solar panels and the regulator 33

Between the batteries and the various connected elements (regulator, converter, charger) 33

Battery bank voltage monitors 34

How low can you safely charge your battery? 36

Fuses and Fuse Holders 36

How to calculate the fuse size 37

Important Locations and Ratings For Fuse 37

Other Power Sources 39

Shore Charging (Plug-in Chargers) 40

Generators 40

Wind Turbines 41

How to install a solar power system 41

How to install a battery bank 42

TOOLS REQUIRED 42

SERIAL CONNECTION OF YOUR BATTERIES 43

PARALLEL CONNECTION OF YOUR BATTERIES 43

How to install a solar Charge Controller 44

Charge controller connection sequence 45

How to install the solar panels 46

Installation of the roofing screen 49

Installation of the sealing system 49

Solar Panel Safety Lines 51

Should You tilt your solar panel 52

How to wire up your solar power system 53

How to crimp 53

To install your roof hatch 57

Adding DC 12 Volt Appliances 61

Adding efficient interior lights to your vehicle 62

Temperature regulation appliances 63

Cooling your vehicle 63

Heating Your Vehicle 63

Other methods 64

How to use a Bulk Dc-Dc converter 65

Adding AC Appliances 68

Off-Grid Internet	68
4G LTE Router with High Gain Antenna	69
Smart Home Appliances	69
Solar System Maintenance Schedule	70
Odds and Ends	72
Phantom Loads	74
How to find Phantom Loads	74
Storing a Solar Power System	75
Connecting different types of solar panels together	76
Connecting different Solar Charge controllers to one battery bank	76
Solar Electric Cooking and food preparation	77
Solar Water Heating	78
Should I install a Battery Isolator?	79

Introduction to Electricity

This modest explanation is for those who are confused or have little knowledge of electricity. As someone who has little or no understanding of electricity, this section will help to give you an idea of what electricity is and how it works.

If you are an expert, you will have to forgive certain simplifications. This is because the aim is to allow everyone to find their way around their electrical installation.

Electricity is a form of energy. It is often described as a phenomenon linked with the mobility or immobility of charged particles. Natural phenomena, such as lightning, were already observed from way back. But for a very long time, electricity terrified men who saw it as a manifestation of divine anger or supernatural powers.

It was not until the late 16th century that it began to be studied by scientists to understand its mechanisms and establish laws. Their successive work made it possible to artificially create electricity by transforming various sources of energy.

Today, this electricity is produced by power plants that transport and distribute it to consumers. Like the fire of the time of prehistoric men, electricity has changed the lives of humanity. It has become essential to everything that makes up our daily life.

What makes up electricity

Electricity just like every matter is made up of atoms. Each atom is made up of:

- A central nucleus which is an assembly of protons and neutrons. Protons carry positive charges and neutrons do not carry charges and are therefore neutral (hence their name);
- A set of electrons which rotate very quickly around the nucleus. Electrons carry negative charges.

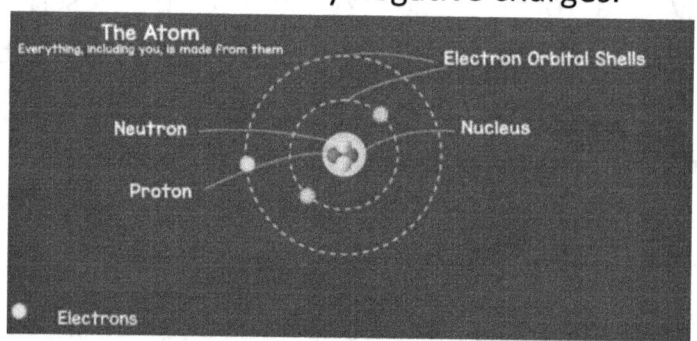

Normally, an atom includes as many electrons as protons, so as many positive charges as negative charges. These charges often balance, making the atom electrically neutral. However, an electron may be added to those of an atom (by friction with another atom for example). This can cause the balance to break and the atom becomes negative. In the same way, an electron can be removed from an atom, causing it to become positive. Electricity results from the movement of these electrons.

The Forms of electricity

Electricity naturally occurs in various forms. These forms include:

- The nerve impulses of certain living organisms which emit electric discharges (as for the eel);
- The static electricity created by rubbing or bringing different materials into contact, for example between hair and wool when removing a sweater;
- Lightning which is formed from an electric discharge between the Earth and a cloud. Sometimes, it occurs only between two clouds.

It can also be created artificially in power plants by transformation:

- Sources of fossil fuels like coal, petroleum or natural gas, resulting from the decomposition in the rock of plants during millions of years;
- Sources of fossil energy such as uranium, the atoms of which can be broken down to release heat and energy;

Renewable energy sources such as water, wind, sun, heat from the Earth, or biomass that nature constantly renews.

Measuring electricity

In terms of measurements, four well-known parameters make it possible to evaluate an electric current. These measurements include voltage, intensity, power, and energy. To understand these physical measures, you can compare the circulation of electrons in a conductor to that of water in a pipe.

The electrical voltage is comparable to the pressure of the water in the pipe when the tap is closed. It is the accumulation of charges in the conductor, which results in the appearance of an electrical potential. Called voltage, this potential is expressed in volts (V).

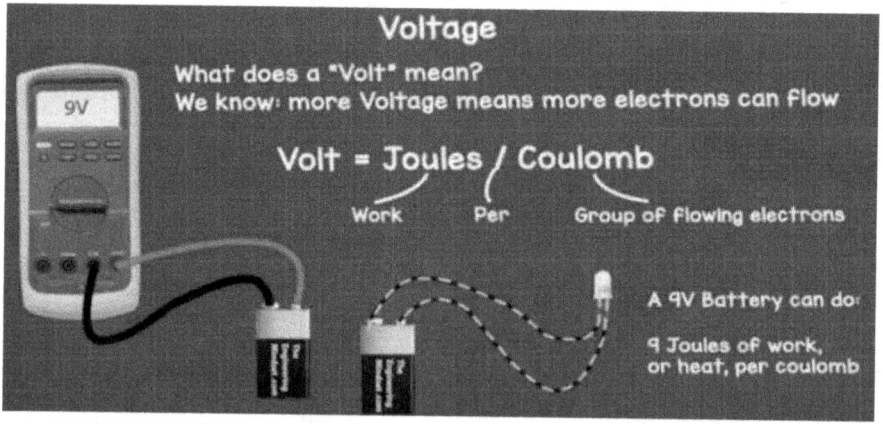

The intensity of the electric current can be compared to the flow of water passing through the pipe when the tap is open. This flow of electrons in the conductor is expressed in amperes (A).

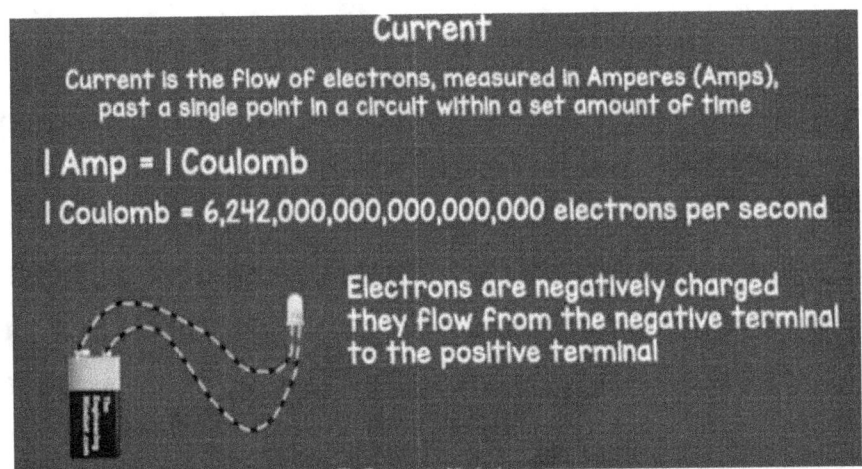

The power of the water jet corresponds to the result of the combined pressure and flow. Likewise, electrical power is the product of "intensity" and "voltage". Its unit of measurement is the watt (W). It is more commonly expressed in kilowatt (kW).

Finally, energy evaluates the consumption of electricity, which is to say the power used during a given period. It is measured in kilowatt-hours, that (kWh).

Electric Circuits: Serial vs Parallel

The electrical circuit can be designed in parallel or series depending on the specificity of the electrical installations connected to it. But in practice, it is the parallel circuits which are used to carry out domestic cabling. An assembly with multiple advantages, this cabling technique is much more appreciated than the series circuit.

The difference between the parallel or serial circuit

Electrical circuits intended to supply various installations with electricity can be designed in two ways, either in parallel or in series. The parallel circuit is a mounting method that supplies currents to its parallel components at the same voltage, but the current is not the same. In other words, it is an electrical distribution technique whose branches share pairs or triple common nodes. This type of circuit is most commonly used in domestic electrical networks.

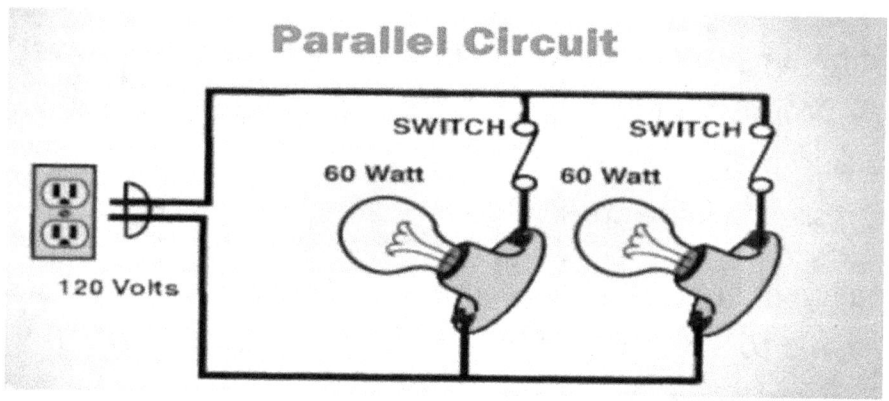

As for the series circuit, it is a principle of mounting cables that allows the different elements of the electrical installation to have an individual load. On this type of circuit, each device or point of use is continuously supplied with a load that is exclusive to it. With a single sequence and a single series, the current flows in one direction until all the elements of the circuit are supplied. The difference between the parallel or series circuit, therefore, lies in the mounting and operating technique.

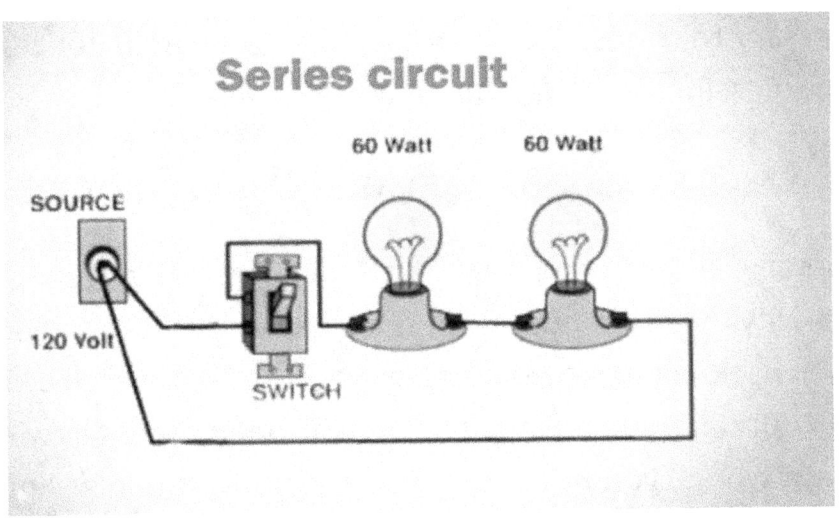

The multiple advantages of the parallel circuit

Between the circuit in parallel or series, most expert electricians recommend the parallel connection for its multiple advantages. Indeed, this type of circuit allows elements that are installed in parallel but remain independent from each other. In parallel circuits, the various electrical components are provided with their switches so that each of them can operate independently.

Another advantage of the parallel circuit is its ability to provide the same voltage as the source. Thus, the multiple bulbs mounted on a circuit will have the same brightness. Also, as required, components can be added to a parallel circuit without changing the voltage. In terms of assembly, the parallel circuit is also simpler to set up while being more secure and reliable. For all these reasons, when it is necessary to choose between the circuit in parallel or series, the professionals opt for wiring using parallel circuits.

Overview of Major Solar Panel System

Generally, a solar panel system is a device that generates electricity when exposed to sunlight. It turns the electric meter back, allowing you to have access to solar electricity both during the and night. To fully understand how this system works, you need to get an idea of what Solar Energy entails.

Solar energy is the energy scattered by the Sun's radiation. From radio waves to gamma rays to visible light, all of these radiations are made up of photons, the fundamental components of light, and the vectors of solar energy. Solar energy comes from the nuclear fusion reactions that animate the Sun.

On Earth, solar energy is the source of the water cycle, the wind, and the photosynthesis of the plant kingdom. The animal kingdom, including humanity, depends on the plants on which all food chains are based.

Solar energy is the source of all forms of energy production currently used on Earth, except for nuclear energy to geothermal energy and tidal energy. Man uses solar energy to transform it into other forms of energy: chemical energy (the food our body uses), kinetic energy, thermal energy, electrical energy, or biomass.

By extension, the expression "solar energy" is often used to designate electricity or thermal energy obtained from the primary energy source that is solar radiation. Currently, there are two main ways of exploiting solar energy:

- Photovoltaic solar which directly transforms solar radiation into electricity;
- Thermal solar which directly transforms radiation into heat. The so-called "thermodynamic" solar is a variant of thermal solar. This technique differs in that it uses the thermal energy of the sun to transform it a second time into electricity.

Many research programs are underway to improve the yields of new technologies for exploiting solar energy.

Photovoltaic Solar System

The term "photovoltaic" designates the physical phenomenon (discovered by Alexandre Edmond Becquerel in 1839), or the associated technique. The advantage of this system is to convert the sun's energy directly into electricity.

Photovoltaic solar energy is the electricity produced by transforming part of the solar radiation in a photovoltaic cell. Photovoltaic cells are made from semiconductor materials, such as silicon, produced from the raw material of very high purity.

Although still accounting for a very small share of global electricity production, solar photovoltaics has a great future. This is as a result the expected progress, lower costs, simplicity, and versatility. Being able to operate with or without connection to a network, it can meet the electrical energy needs of a house (sensors on the roof) or an industry.

The first application appeared in the space domain for satellites. Other sectors then used photovoltaic technology, notably telecommunications, maritime and aerial beaconing, domestic lighting and water pumping. But as the sun is not visible at any point of the Earth 24 hours a day, this application requires the use of batteries or other systems which ensure the storage of electricity for consumption outside the period of sunshine.

Components of a solar power system

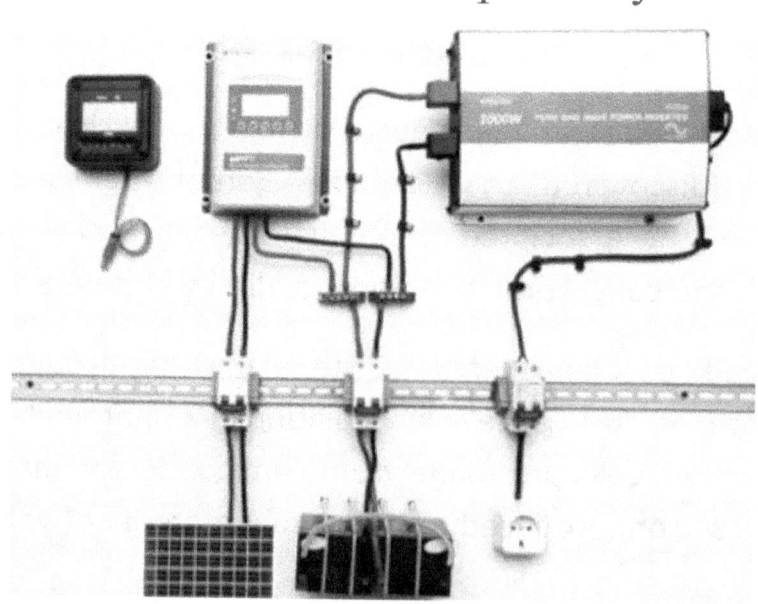

From the overview above, it is evident that a solar panel alone is not enough. For example, a pump connected directly to a solar panel will only work when the panel is exposed to the sun. Therefore, different components help to make this system work properly.

The regulator

This is the essential element for any solar installation, it will optimally charge your batteries while preserving them from a possible overcharge or complete discharge. Depending on the model, it can also manage automatic lighting on and off.

There are two main families of solar regulators; PWM regulators and MPPT regulators. The solar panel (s) are connected to the regulator, the battery bank)which for small solutions can be limited to a single battery), the equipment to be operated (lighting, pump, alarm, road signs, weather equipment, rural electrification, electronic equipment, etc.). Only a system using an oversized battery bank with a voltage that is equal to the maximum voltage of the solar panel and discharge within the critical discharge of the batteries can do without a regulator.

The battery

There is no other known way to store energy and restore constantly or on-demand than a battery. Apart from injection solutions connected to the EDF network, the batteries are therefore insensitive to any autonomous solar installation.

However, do not think that any battery will do. As the charge and discharge cycles are constant, it is necessary to use suitable batteries whose life cycle (charge/discharge) and the discharge capacity will be as large as possible while having good temperature resistance.

The converter or inverter (optional)

If the equipment to be connected is at low voltage and corresponds to the battery bank voltage, the use of a converter will not be necessary. If the equipment is 230V AC, it will be necessary to use a converter or inverter capable of converting the battery voltage into the main voltage. The converter needs to have a power that is at least 30% higher than the actual need of the application. Also, different types of converters will provide a waveform more or less at the voltage supplied by the EDF network.

How Does the System work?

The silicon wafers laminated under anti-reflective glass in the solar panels collect the photons from the sun and transform that into continuous electrical energy. This energy flow, then to an inverter, which converts and transforms it into usable voltage and alternative electrical energy.

Solar Panel System Design Methods

The design of a solar panel system generates many questions that must be considered before diving into the implementation. For example, it will be necessary if your solar electric or photovoltaic van is the right size or the costs of installation and maintenance. Also, all these important points have to be treated with meticulousness to optimize energy efficiency. Consequently, in addition to contributing to the maintenance of the good health of our planet, the type of installation indicates a certain monetary interest.

Fast Method

Lazy method

The lazy method of selecting a solar power design involves putting certain features into consideration. Some of these features include budget, weight, durability e.t.c.

The minimalist

The minimalist design is a good option if you wish to install the system off-road (car or minivan) with a limited roofing space. It also offers an aerodynamic and lightweight setup. However, you may not be able to run large appliances like microwaves and large motors.

The classic 400 watt

The power or wattage is one very important feature to consider when choosing a solar power design. The classic 400-watt solar power system offers 400 watts panels which are the average wattage you can get from solar panels today. Despite the solar projects that you may be considering, a 400-watt solar system is a great option.

The off Grid king

The off-grid system term refers to the system not relating to the grid facility. The off-grid system is also called a standalone system or mini-grid which can generate the power and run the appliances by itself. Off-grid systems are suitable for the electrification of small communities. Electrical energy in the off-grid system produced through the Solar photovoltaic panels needs to be stored or saved. This is because requirements from the load can be different from the solar panel output. Better banks are also used for the purpose generally.

Ultra-lightweight

Do not ignore these criteria when choosing a design for your mobile solar power system. Indeed, for a good quality material, which will assure you a supply of energy at all times and in all hours, you must opt for a powerful material. But this material still has to be light enough since the system is expected to be mobile. For a nice rendering, you can choose the solar panel kit according to the design of your motorhome.

Low budget

When selecting a design for your mobile power system, you may need to pay attention to the cost as well. It is essential to find a low budget design that will suit your power needs while still offering other required features.

Dystopian future

While solar panels designed for permanent installation on a single site maybe a little more fragile, mobile solar panels must be much more durable. Pay attention not only to the quality of the panels but also to the quality of the frame, materials, and the quality of cables, clamps, and connectors. You need robust durability in all of the components of your portable motorhome solar panel kit, just like you need it for all of your other equipment.

Traditional Method

In terms of the traditional method for solar power systems, a lot of calculations are considered. Below are some calculations you may need to put into account when getting a solar power system.

Calculating the load

It is defined according to the battery (voltage), the panel, and you, the consumer.

A 90 WC panel in 36 cells, for example, will provide a maximum current of 5.5 amperes under nominal conditions. It has a count of 20% of margin for the intensity admitted by the regulator in an entry, 40% if you are by the sea, or in the mountains, and especially in winter. This is because the lighting is more important with the 'nominal' conditions.

So with 40% margin, 5.5 amps / (1-40 / 100) = 5.5 / 0.6 = 9.2 amps. A 10 amp regulator is therefore sufficient at the input.

As consumers, we often have a total of 5 watts of light bulb + 50 watts of TV, so if everything works at the same time, we consume 55 watts.

Generally, $P = UI$, so $I = P / U = 55/12 = 4.58$ Amperes. A 5 amp regulator is therefore sufficient for the output.

In general, the regulators manage the same intensity as input and output. It is therefore necessary to align with the highest value in terms of load.

Calculating the battery bank size

A battery is defined by its voltage in volts and its 'capacity' in ampere-hours. The energy it contains is its capacity multiplied by its voltage:

E = U (volts) x capacity (Ixh), i.e. for our example: 12 volts x 60 Ah = 720 volts x (Amps x hours) = 720 (volts x amps) x hours = 720 watt x hour = 720 wh.

The life of the battery is calculated in cycles. The deeper a battery is discharged, the fewer cycles it will be able to provide.

The ideal is not to discharge its battery of more than 10% of its capacity, we consider that it is not cycled and does not (almost) age. But it is expensive and it weighs heavily (I think of motorhomes). So we agree to discharge it from 20 to 30%, it's reasonable and the number of cycles, depending on the technology, remains quite high. For example, for an inexpensive, open lead type battery, of the Enersol range, at 20% discharge, we find ourselves substantially at 400 cycles.

Some professionals go as far as to discharge the battery up to 70%. Occasionally, this is acceptable, but in good working order, this amounts to depleting the battery very quickly.

Calculating Solar Array Size

You can calculate your solar array size by first of all calculating the output you need. This can be done by dividing your daily output (kWh) by the peak sun-hours. You can then divide your output by the efficiency of each panel. This will give you the number of panels you need for your system.

Solar array size = required power (Wp) / power per panel (Wp)

Example (household of 4 people): 4,706 Wp / 250 Wp = 19 solar panels

How to calculate the maximum/ minimum solar array size for a battery

To go from a figure in kWh to several solar array sizes for a battery, multiply by 1000 (to convert kWh to Wh) then divide by the voltage at the battery terminals (to convert Wh to Ah) then by the capacity of the batteries (in Ah) and round up.

Example:

We use batteries whose voltage is 12V and the capacity 200Ah:

9.16kWh x 1000 = 9160Wh

9160/12 = 763Ah

763Ah / 200Ah = 3.81

So, we need 4 batteries

Other solar array sizing tips

Determine the type of solar panel you wish to use

There are a variety of solar panels one can choose from. These solar panels range from large 250-watt panels to the small 20 watt panels at volts of 12, 24, or 48V. The type of solar panel you choose plays a huge role in sizing and designing your solar power system.

Calculating Solar Charge Controller size

To find the correct minimum size for your solar charge controller, you can use the basic guide below, depending on whether you want to use a Pulse Width modulation regulator (PWM) or a maximum power point tracking regulator (MPPT). Note: this calculation assumes that you are using solar panels of the same type.

Efficiency Considerations

In terms of efficiency, there are a few things one must consider:

1) Combined power of a solar system

For example. two panels of 150W = 300W combined power of the system

2) Rated voltage of the solar system

Solar panels usually have a rated voltage of 12V or 24V. if you connect them in parallel, they will remain at the same rated voltage. If they are connected in series, the Voltage will increase.

Example: two panels 150W 12V in parallel = 12V rated system voltage

Example: two panels 150W 12V in series = rated voltage 24V of the system

Other Factors To Consider

1) Maximum power voltage of the solar system

This voltage is different from the rated voltage. Typically, a 12V panel would have a maximum power voltage of about 18V, while a 24V panel would have a voltage between 30V and 36V. The same rules as above apply depending on whether your panels are in Series or parallel.

Example: two 150W panels with a maximum supply voltage of 18V, in parallel = 18V maximum supply voltage of the system

Example: two 150W panels with a maximum supply voltage of 18V, in series = 36V maximum supply voltage of the system

2) Battery charging voltage

This is the voltage at which your batteries will be charged. Different types of batteries charge at different voltages, check the specifications of your batteries if in doubt.

Example: 1X 12V AGM battery would be about 14.4 V

Example: 2X 12V AGM batteries in parallel would be about 14.4 V

Example: 2X 12V AGM the batteries in series would be about 28.8 V

How To Select Solar Power System Components

When selecting the components for your solar power system, like battery, panel, or charge controllers, there are few things you should consider. The features to consider are often specific to each component. In this section, we will break down each component to help make your selection easier than expected.

Selecting a Battery

Several storage systems are sold on the market. Each type of battery has its advantages and disadvantages. Therefore, the choice will depend on the use for which the battery is intended.

Lithium-Ion batteries

Lithium-Ion batteries are the best known because they are present everywhere today. This type of battery is used in electric cars, smartphones, or even laptops.

Its advantages are manifold. It does not require maintenance and has good longevity. Also, it is non-toxic and can be recycled. The main negativity is its high cost compared to other energy storage solutions.

Open lead batteries

This type of battery works with lead. They are said to be "open lead" because their functioning releases oxygen and hydrogen. This requires placing them in a well-ventilated area.

They are quite resistant when they are well maintained. Their lifespan is more than 10 years. However, they are not waterproof, remain very sensitive to cold, and can freeze.

AGM batteries

It is a kind of lead battery. It has several advantages over that of open lead. AGM batteries are waterproof, so they do not produce heat. Also, they do not require maintenance.

However, they have a lower number of charge-discharge cycles. Indeed, in use, a battery is subjected to charges then successive discharges. This is called cycling. Certain types of battery support this effort more or less. Thus, AGM batteries have relatively low cycling compared to lithium-ion batteries for example.

Gel batteries

This type of lead battery is the most successful. Like the AGM battery, the gel battery is waterproof and does not require maintenance. It is frost resistant and has an excellent lifespan. However, it remains sensitive in the case of excessive loads.

The main Technical Characteristics of Batteries For Solar Power Systems

The installation of solar panels with solar batteries has specific material needs. Thus, one cannot compare the electric battery of a car with a battery for photovoltaic panels. This is the reason why there are batteries specifically designed to work with solar panels.

Solar energy will charge the batteries sporadically. Thus, the charge and discharge cycles will not be complete and the energy may remain in the battery for some time.

Things To consider

In the case of electrical installation at home, it is necessary to take as little loss as possible to take full advantage of the gain offered by the self-production of energy. Thus, the batteries must have optimal charging capacities. This is called good charge efficiency.

Let's explain this point. When a battery is charged with energy, ideally, we try to recover all of the stored energy. In this case, the yield would be 100%. Unfortunately, this is not the case, because part of the energy dissipates.

Lithium batteries have a charge efficiency of around 90%. In other words, 90% of the energy that was previously stored in the battery is recovered during use. You could say that they have good charge efficiency.

This charge must be rapid to gain efficiency, and this stored energy must be conserved. This is the reason why a good battery should have the lowest possible discharge rate. Concretely, the batteries marked C100 take 100 hours to discharge while the C5 will take only 5 hours to discharge.

Selecting Solar Panels

There are currently 3 main types of solar panels, which are differentiated by the type of cells that compose them. All the cells are produced based on silicon, but the different manufacturing methods give them very different characteristics, in particular in terms of productivity. The modules typically have 36, 54, 60, or 72 cells.

For information: a monocrystalline silicon cell is always octagonal (or round), while a polycrystalline cell is always square.

Monocrystalline cells

Monocrystalline cells come from a single block of molten silicon, so they are very "pure". They offer the best yield (between 13 and 17%) but are also more expensive for production, therefore for sale. These cells are generally octagonal and of a uniform dark color (navy blue or gray). These cells are the most efficient, they therefore make it possible to constitute panels that have better yields, which produce the most energy with the least surface.

Polycrystalline cells

Polycrystalline cells are made from a block of silicon crystallized in the form of multiple crystals. When viewed up close, you can see the different orientations of the crystals. They have a yield of 11 to 15%, but their production cost is lower than monocrystalline cells. They are generally rectangular and are midnight blue with reflections.

Compared to monocrystalline panels, their performance is even worse when the sunshine is low.

Amorphous cells

Amorphous cells are produced from a "silicon gas", which is sprayed onto glass, flexible plastic or metal, by a vacuum vaporization process.

The cell is very dark gray. It is the cell of calculators and so-called "solar" watches because this type of cell is inexpensive and the technology can be used on many supports, in particular flexible supports. The problem is that its yield is 2 to 3 times lower than monocrystalline cells.

Flexible Solar Panel

The flexible photovoltaic panel allows you to generate energy wherever you are. It is perfectly suited to an electrical installation for motorhome, caravan, or even boat. Light and flexible, you will have no trouble installing and connecting it. The panel is available in 50W, 110W, and 170W power. With a flexible panel, you benefit from high-performance Maxeon solar cells, resistant to corrosion, cracking, and harsh weather conditions.

Installation of the flexible solar panel

The flexible solar panel fits perfectly into a boat solar kit, solar kit camper, or a solar kit caravan. You can install it wherever you want, it just requires that the panel is well exposed to maximize its production. You can install the panel with stainless steel gourmets and/or with adhesives. The flexible solar panel is fitted with standard quick connect cables.

Selecting a Solar Charge Controller

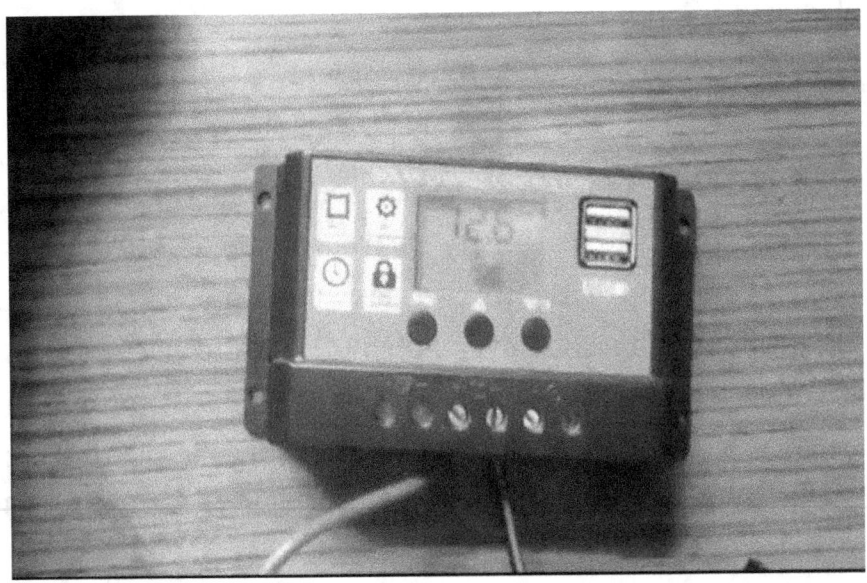

The solar charge controller is the central element of a stand-alone system. It controls the flow of energy throughout the system and is essential for operation and lifespan. This is why it is important to select the proper solar charge controller for your solar power system. The cost of the solar charge controller is only 3 to 5% compared to the total cost of a stand-alone system.

However, it remains the most important element of the system. Purchasing a high quality and reliable solar charge regulator from a high price range pays off very quickly, as it significantly contributes to improving the long battery life and therefore reducing the cost total of the system in a big way.

To choose your solar regulator, you must take into account 3 main elements:

- The maximum open-circuit voltage of your panels: The regulators accept more or less wide voltage ranges.
- The minimum voltage for the batteries.
- The maximum intensity of the regulator: the intensity of the regulator (expressed in amperes) must be greater than the intensity of the short circuit of the solar panels to which it is connected. This information is indicated on the instructions for each panel. It is preferable to take also a safety margin of 10% to 20%.

MPPT Solar Charge Controller

The MPPT solar regulators (for Maximum Power Point Tracking) is a new technology that obtains the best performance from photovoltaic solar panels, from 10% to 25% more energy.

An MPPT solar regulator has a scanning function, which scans the voltage of the solar panel every two hours to find the maximum power output point. This allows it to adapt the voltage delivered by the panel to that which the solar batteries can absorb. These controllers not only increase the energy production of a photovoltaic installation but by optimizing the charge of the batteries, they also significantly extend the lifespan of these batteries. The MPPT regulator, more expensive than these PWM colleagues, will be able to modify the module voltage and provide the voltage necessary to charge 100% of the batteries.

Selecting an inverter

To select your solar inverter correctly, you will have to ask yourself the question of the battery pack voltage, the supply voltage of the receivers, and the power of the receivers.

Battery voltage

This is the DC voltage of the solar installation. In volts, it can be 12, 24, or 48 volts. You can also have 36 and 96 volts but this is a bit rare. The larger your solar panel installation, the higher the voltage.

Receivers supply voltage

In a conventional home, the receivers are powered by a voltage 230. In some industries, however, the voltage will rise to 400 to supply the machines with electricity.

Inverter power

This is expressed in volt-amps (VA) or Watt (W). It is the power that the inverter can release at a constant speed.

The different models of solar inverter

Once you have this information, you could choose your solar inverter. There are three models.

The module inverter

It is a small model hanging behind the photovoltaic panel. The latter then directly produces 230 VAC. With the module inverter, you will benefit from wiring limited to the alternative, a simple binding to the house, and a reduced sensitivity to possible shading.

The chain inverter

This is connected to each chain of solar panels in series. It works with greater power and greater voltage than the module inverter. This also allows a better yield.

The central inverter

It is the most used model for homes and villas. Its particularity is that it separates the part of the direct current from that of the alternating current. On the other hand, the wiring is substantial and bulky with a strong sensitivity to shading.

Selecting wire

It is imperative to choose the right cable diameter for your photovoltaic installation: In fact, fire risks can occur for an undersized cable section (cable overheating). Also, making the right choice before installation allows you to avoid online losses and additional costs due to the purchase of an oversized section.

The diameter of the cables is determined according to the intensity of the current (A) and the distance to be covered. Cables with a large cross-section for direct current transmission over long distances are expensive. It is economically wise to make the correct calculation of the section to be ordered before installing the photovoltaic power plant.

Tips: Increasing the nominal voltage of the installation (from 12V to 24V or 48V) makes it possible to reduce the diameter of the cables required and thus the cost of the installation. Indeed, if for example the current is 20A at 12V (U x I = P = 240W), in 24V the current passing, for the same power, is divided by two, that is 10A (24 x 10 = 240W). The same goes for a system voltage of 48V: The current passing is 5A (5 x 48 = 240W).

You can quickly determine with the chart below the cable section to be used according to two variables: The maximum passing current (Amperes) and the cable length (meters).

For example, for a current of 12A and a length of 14 meters, the cable must have a section of at least 4 mm² in diameter (see abacus above).

For example, for a current of 12A and a length of 14 meters, the cable must have a section of at least 4 mm² in diameter (see abacus above).

You can also determine the cable section suitable for your installation using the tables below:

12 Volt Wire Gauge Chart

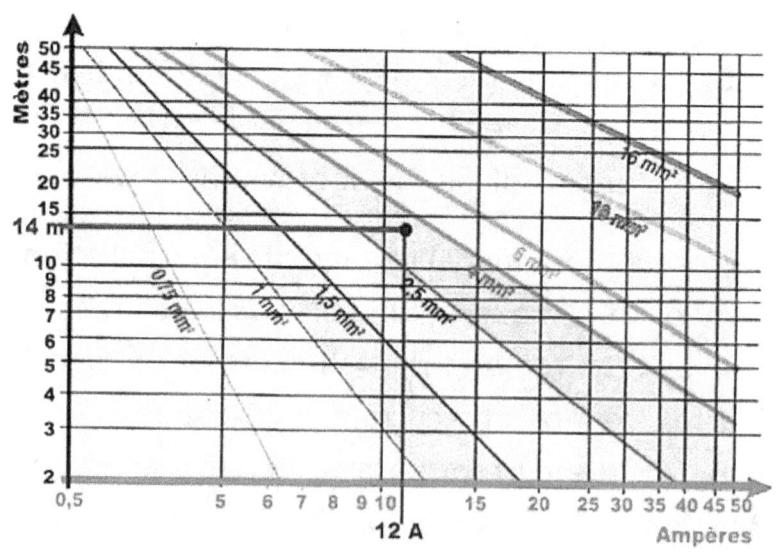

Between the solar panels and the regulator

Cable section	2.5m	5m	7.5m	10m
0.75 mm²	3.4 A	1.6 A	1.2 A	0.9A
1.5 mm²	6.7 A	3.4 A	2.2 A	1.6 A
2.5 mm²	11.2 A	5.7 A	3.5 A	2.8 A
4 mm²	18 A	9 A	6 A	4.5 A
6 mm²	27 A	13.5 A	9 A	7.5 A
10 mm²	45 A	22.5 A	15 A	12 A
16 mm²	72 A	36 A	24 A	18 A
25 mm²	112.5 A	57 A	37.5 A	28.5 A
35 mm²	157.5 A	79.5 A	52.5 A	39 A
50 mm²	225 A	112.5 A	75 A	57 A

Between the batteries and the various connected elements (regulator, converter, charger)

Cable section	2.5m	5m	7.5m	10m

0.75 mm²	2.3 A	1.1 A	0.8 A	0.6 A
1.5 mm²	4.5 A	2.3 A	1.5 A	1.1 A
2.5 mm²	7.4 A	3.8 A	2.5 A	1.9 A
4 mm²	12 A	6 A	4 A	3 A
6 mm²	18 A	9 A	6 A	5 A
10 mm²	30 A	15 A	10 A	8 A
16 mm²	48 A	24 A	16 A	12 A
25 mm²	75 A	38 A	25 A	19 A
35 mm²	105 A	53 A	35 A	26 A
50 mm²	150 A	75 A	50 A	38 A

Battery bank voltage monitors

The objective of a voltage monitor is to provide all the information on the energy of the battery that is produced and consumed. It also helps to deduce the condition of the batteries.

The charging current, which is injected into the battery is a variable element, it takes into account the capacity of the battery and its state. It becomes even more complex when there are several sources providing current at the same time like solar panels etc.

The voltage monitor must take all these parameters into account to be able to display the essential data: voltage, current (product, and consumed). It must also be able to make the difference and deduce the actual current flowing in or leaving the batteries as well as the actual capacity remaining in the batteries.

Complex? Yes or even impossible if we had to do the calculations manually. But, with these electronic tools, this does not pose any difficulty if one takes the time to configure them properly. Indeed, several important points will have to be taken into account for the good programming of the manager. This will take into account the CEF yield and the Peukert index. The CEF is the charge efficiency coefficient which takes account of the age of the batteries. The Peukert index on the other determines the apparent capacity of the battery. These last two points will be the basis of an exact reading of the state of the battery park.

For wiring, it is very simple because, in practice, a resistance of low value and high precision is placed between the batteries and the electrical circuits. The passage of current in this shunt causes a small drop in voltage. This voltage is sent to the controller which constantly calculates the difference between the current entering the battery and the outgoing current. The latter then displays all the information.

How low can you safely charge your battery?

The safest low Voltage at which a battery can be permanently maintained to be sure that it is charged when it needs it is 13.5 V at 13.68 V at 25 ° C. This value should be corrected by 0.005V more or less per centigrade degree depending on whether the temperature drops or rises. At -10 ° C it is 14.16V and at + 40 ° C 13.26V.

Fuses and Fuse Holders

Fuses come in the form of a cylinder (ceramic or glass) and contain a conductive filament intended to melt if the expected intensity is exceeded over a certain period.

There are fuses of different intensities (in amps) to effectively protect each electrical circuit in the house:

2 A, for the peak / off-peak contactor

10 A commonly used to protect lights or a group of electrical sockets

16 A, more powerful, used for certain devices

32A, The most powerful, used to protect an oven or a washing machine for example.

In case of overcurrent (electrical overload due to too many devices operating on the same fuse), fault or electrical short circuit, the filament of the fuses blows due to the increase in temperature, allowing the circuit to be cut and therefore protect the devices connected to it.

Unlike the circuit breaker, which simply needs to be reset when it blows, the blown fuse must be replaced.

If you have a fuse distribution board, it is therefore recommended to have one at home to quickly resolve the problem. It is much cheaper than the modern circuit breaker and is also easier to change.

How to calculate the fuse size

To size a fuse, you just need to know the voltage (in volts) and the intensity (in amps) delivered by the equipment that you want to protect.

The general calculation is as follows: Current of fuse = Power of the device / Voltage of the device.

The power and the voltage must be taken under the maximum conditions, which you will find in the description of your product.

Important Locations and Ratings For Fuse

There are four important locations where fuses are placed. Below are the locations and the rating for each of them.

1. Between the solar panel and the regulator

Between the panel and the regulator, we speak more of a circuit breaker or a protective box. This helps to decouple the panel from the installation if operations are to be expected or if the system is intended to be removable. This kind of protection is done on large solar installations and rarely on vehicles. Indeed, it is quite simple to disconnect your solar panel from the regulator if an operation is to be expected.

To choose the intensity of the circuit breaker or fuse, here is the operation to perform:

Fuse current = (Panel power / Panel voltage)

The power and the voltage of the panel must be taken under the maximum conditions, which you will find in the description of your product.

2. Between regulator and battery

For the fuse between the regulator and the battery, it is the same principle in simpler terms. This is because we already have the intensity of the regulator.

Example:

A regulator has a maximum output current of 15 A.

It is therefore just necessary to choose the higher rating, that is 20 A, supporting the voltage of the 12v or 24v battery bank, depending on the choice you have made for your battery.

3. Between the battery and the converter-charger

For the fuse between the battery and the converter-charger, the calculation given at the beginning should be carried out:

The current of the fuse = Power of the converter-charger / Voltage of the charger converter

4. Between the auxiliary battery and the coupler-separator

For the fuse between the auxiliary battery and the coupler-separator, it is very simple because the amperage is given on the product.

For example: Concerning the battery coupler with a 12 / 24V 120A Battery combiner kit, its amperage is 120 amps. So the fuse should have a current of 125 A and support a minimum current of 12 or 24 volts depending on your battery.

Other Power Sources

Because it exists in an unlimited quantity on the scale of human time, it is said that solar energy is renewable energy. But there are several other sources of power that can be used to generate electricity or heat.

Shore Charging (Plug-in Chargers)

Shore charging, which is also known as shore power or an alternative source of power, refers to the technique that provides a van with electrical power in campgrounds with plugins. This allows the shutting down of the generators to stop the diesel consumption of the van. Shore charging is therefore an effective way to charge up your batteries during the winter when the sun isn't out for long.

Generators

Generators use fossil fuels to generate electricity. Fossil fuels are naturally present in the subsurface of the Earth. They are formed from organic matter that decomposed over millions of years. They exist in three forms: petroleum, natural gas, and coal. They are used as fuels, as fuels but are also used to produce electricity.

Generators are currently the most used energy source in the world (80%). They are often singled out for their polluting nature, but the international exploitation of oil, and coal to a lesser extent, does not stop.

The gas and oil deposits are located mainly under the seas and oceans. To recover them, it is necessary to practice drilling. As for coal, it is in the basement and requires digging mines to extract it.

These substances are hydrocarbons: they are composed of hydrogen and carbon which make them substances with high energy power. It is their combustion that will allow the production of electricity in thermal power plants.

Wind Turbines

Among the various known renewable energies, wind energy is that of the wind. It corresponds to the movements of masses of atmospheric gases essentially driven by the rotation of the Earth and temperature differences on the surface of the globe. The kinetic energy of these winds can then be converted into electricity using wind turbines.

How to install a solar power system

A Solar Power system involves a sensor and the presence of several elements (depending on whether the system is autonomous or not) that make up this system. These elements include solar cells, batteries, charge regulators, DC/AC converter.

Solar cells (or PV), the basic elements of these modular systems, make it possible to directly convert solar radiation into electrical energy. The cells are assembled in modules or solar panels, and these can form a field. The module is the best known and most popular support and it is often installed directly on the roof of houses. But there are many other applications. It is indeed possible to install the photovoltaic modules on buildings, to incorporate them into the design of a structure.

How to install a battery bank

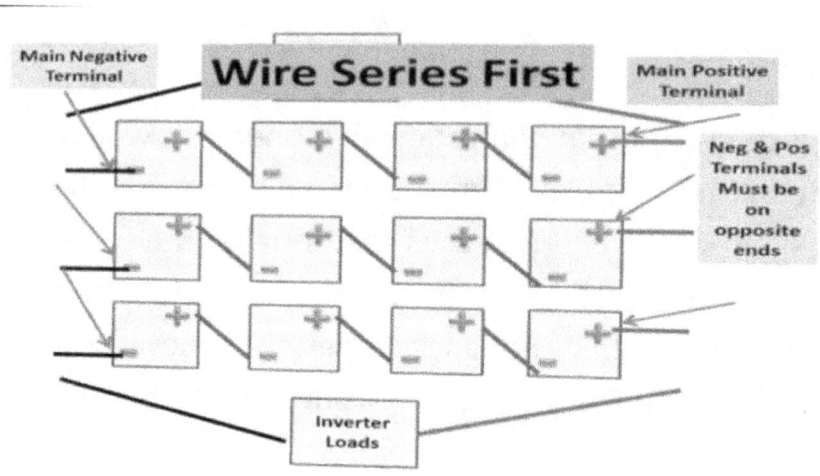

There are currently two main methods of installing a Battery bank in a solar power system. These two methods are the series and the parallel method. This guide to understanding how to mount your batteries in series or parallel.

TOOLS REQUIRED

- 2 or more batteries
- Flat screwdriver
- Pliers (for crimping)
- A key of 8
- Stripping pliers

SERIAL CONNECTION OF YOUR BATTERIES

The series connection requires you to connect the pole (+) of one battery to the pole (-) of another.

Example:

With 2 batteries of 12V 100Ah, we will obtain a battery of 24V 100Ah.

With 4 batteries of 12V 100Ah, we will obtain a battery of 48V 100Ah.

PARALLEL CONNECTION OF YOUR BATTERIES

Parallel mounting consists of connecting the plus (+) poles and the minus (-) poles.

Example:

With 2 batteries of 12V 100Ah, we will obtain a battery of 12V 200Ah.

With 4 batteries of 12V 100Ah, we will obtain a battery of 12V 400Ah.

Battery connection in series and parallel

Now you can connect your battery bank to the other devices making up your solar kit (charge regulator, voltage converter, inverter, etc.)

However, Parallel mounting is not recommended on partially charged batteries. Make sure that your batteries are identical (brand, reference, amperage, etc.), purchased at the same time, and not discharged before performing a serial or parallel operation.

How to install a solar Charge Controller

You can create a homemade version with your own hands and customize it if you take into account all of them certain recommendations.

It should be noted that when connecting each type of device, it is necessary to use the most suitable type of solar panel. For example, when using a device designed for an input voltage of about 100 volts, you must use solar panels, which have a similar figure to the output which corresponds to this value.

Before connecting the device, you need to choose the most suitable place to install it. The best solution to this problem is a dry, well-ventilated room. It is strictly not recommended to place flammable materials near the device. Also, it is unacceptable that the device is in the vicinity of various sources of vibration, humidity, as well as many heaters and stoves. The place where the device is to be placed must be reliably protected against various atmospheric precipitation and direct sunlight.

Charge controller connection sequence

To get the maximum effect from using such a device, it is necessary to follow the instructions to the letter, as well as to follow a certain order when connecting the device. The process of connecting the charge controller and various devices will not cause many difficulties.

Each model is fitted with special labelled terminals.

Connection of peripherals must be carried out in strict accordance with the symbols on the contact terminals:

it is necessary to connect the battery and the battery using a special wire and a terminal, scrupulously respecting the polarity;

A fuse must be connected to a specific positive wire to protect the instrument;

on the appropriate contacts of the controller, fix the special conductors starting from the battery of the solar panel and scrupulously respect the polarity;

A special lamp must be connected to the specific outputs of the device to control the corresponding voltage.

Caution: Do not violate the specified sequence. For example, it is strongly recommended not to connect solar panels to the controller when the battery is disconnected. This may cause the device to fail. The design of the inverter must be connected to the battery with special terminals.

How to install the solar panels

To install solar panels, you must necessarily be comfortable with working with heights, roofing, and waterproofing to avoid water leaks. You will have to make sure of the good fixing of the panels to be able to resist harsh climatic conditions. This is done by installing hooks, rails, or battens. Depending on the solar panels and their method of attachment, you may need to install an under-roof screen or a waterproofing system. A good physical aptitude is also necessary to intervene on a roof and handle the voltaic panels.

Basic steps

- deteriorate at the location of the solar panel
- Install the covers (lower and side)
- Install under-roof screen or waterproofing system
- Lay the rails and panels
- Connect the inverter to the grid

Tools required

- Screwdriver and/or drill driver with sockets and / orbits
- Circular saw

- Power extension cord
- Metre rule
- Scaffolding or ladder
- Roofer's ladder (depending on coverage)
- Key flat or bushings
- Screws
- Fishplate and other hardware for installing the rails
- Aberration or zinc
- Lead strip
- Battens
- Extruder gun with silicone glue
- Electric wire with IRL tubes and / or bridges if necessary (depending on configuration)
- Protective gloves
- Anti-fall system
- Protective glasses
- Safety shoes

1. Deteriorate at the location of the solar panel

There are two distinct cases. If the roof is covered with mechanical tiles or slate, you will have to carry out a partial removal. You must start by delimiting the precise location where you want to install the solar panels. It is also necessary to remove the tiles on a surface slightly larger than the surface of the panels to be laid. Once this is done, a hole appears in your roof: the photovoltaic panels will be housed thereafter meticulous preparation.

Clean the area with protective gloves because the tiles are abrasive and can injure your hands. Work on the roof should be done with a fall arrest system (net fall arrest or skirts with harness). If you are decupling near the ridge, be careful not to break the ridge tiles or the half-posts. If your roof is not planked (truss structure), do not walk in the middle of the battens or on the under-roof screen as there is a risk of falling.

2. Install the lower and side coverings

The arguments are connecting parts between the roofing materials of your roof and the various elements which protrude from the roof (the solar panels and their fixings). There adaptations specific to each model of mechanical tiles, both for the shape and the color. For slates, it is a slate-colored sheet or an old-fashioned piece of zinc.

The installation consists of a screwed fixing placed on the existing battens or added battens. The use of a screwdriver and a meter rule is necessary to report the exact measurements. The hooks are also fixed with screws.

If necessary, add a batten along the length of the panel location and behind the first row of tiles to gain height and provide better seating for the submergence. The flashing can be made of plastic and delivered with the kit. However, it must be a lead strip that should be unrolled on the previously installed battens and the first row of tiles. The lead strip (of variable width) is glued with silicone glue applied with an extruder gun. The lead band, deformed by hand, follows the shape of the tiles. The boards and battens are cut with a circular saw.

3. Install the under-roof screen or a waterproofing system

Installation of the roofing screen

When installing a roof, an under-roof screen must be installed. The latter is a technical plastic film which has the function of creating a barrier impermeable to water but not to air. It is recommended when installing this under-roof screen to provide an air space between this screen and the insulating complex. If this is impossible and it rests on the insulation, then opt for a screen with high permeability to water vapor or HPV. This allows the passage of vapor and reduces the risk of condensation in the insulating complex. If condensation occurs in the insulation, it becomes less effective, and may even rot. The strips must be covered with the adhesives recommended by the manufacturer and according to its recommendations.

Installation of the sealing system

Depending on the installation, the solar panels can rest on a flexible plastic waterproofing system. This is to be installed on the roof (a rigid foundation made of battens is then necessary). These sealing systems are then screwed onto the battens at the location of the panels.

4. Lay the rails and panels

The rails must be fixed securely to the rafters using hooks (use the screwdriver and suitable screws for this). The solar panels are heavy and offer a great catch in the wind. Complex shapes with possibly certain adjustments adapt their shape to certain tiles allows fixing through the cover. Full derailing is no longer necessary since the structure of the rails is above the cover.

It is particularly important to respect the maximum distance between two rails recommended by the manufacturer as well as the maximum distance between two fasteners to the rafters. If the rails are not long enough or need to be crossed, there are fishplates as well as many other connecting pieces. Note that some solar panels can rotate a quarter of a turn with their long side being found vertically. It may then also be necessary to lay the rails vertically.

The panels are then fixed to the rails using hooks or bolts. Once all the panels are in place, you can fix the upper cover and then install the missing tiles or slates. When installing solar panels on a waterproofing system, the support rails can be directly installed on it.

5. Connect the panels to the inverter

The inverter is an electronic device that transforms the unstable direct current (output of the photovoltaic panels) into an alternating current of 220 V and a frequency of 50 Hz (formerly ERDF). This inverter is itself connected on one hand to the panels, and on the other hand to the non-consumption meter. This meter is found at the head of all the individual electricity production installations connected to the network. It is imperative to know how to read an electrical diagram to carry out this step properly.

Some panels must be connected in series, while others must be connected in parallel. This is to obtain voltage and current values around the nominal values expected by the inverter to preserve its longevity. Two panels of identical characteristics are connected in series if the positive terminal (+) of one is connected to the negative terminal (-) of the other; their tensions then add up.

If each panel produces a voltage of 12 V, the two panels in series produce a voltage of 24 V. If you connect the terminals of the same polarity, the panels are connected in parallel. This assembly produces a voltage of 12 V, but the intensity produced is the sum of the intensity produced by each panel. However, if you connect two groups of panels in series or parallel, the tensions of each group must be equal to preserve the panels and maximize their production.

Solar Panel Safety Lines

A safety line is very important irrespective of how you decide to install your solar panel system. A solar panel safety line is a strong and sturdy rope that connects each solar panel. The safety lines can take up to 20 minutes to be properly constructed but can save a life. Without the use of a safety line to connect the solar panel, there is a high chance of a catastrophic failure.

Should You tilt your solar panel

In theory, the best tilt of the solar panel is the angle that allows the collectors to be perpendicular to the sun's rays. However, the sun moves during the seasons and the day. The radiation at 90° on the fixed surface of the solar panel cannot be maintained unless you are equipped with a device that follows the course of the sun. It is recommended to tilt the solar panels between 15 and 35°. That is, the solar panel must be tilted to form an angle between 15 and 35° with the ground.

In winter, the sun is low compared to the horizon. To capture the maximum light, the panel must be tilted at an angle of about 60° which is almost vertical. But in summer, tilting the solar panel at 60° minimizes the volume of solar electricity production.

In summer, the sun is high. On the contrary, the surface of the solar panel will be perpendicular to the radiation if its inclination is close to the horizontal, i.e. around 10°. However, the inclination of the solar panel at 10° in winter is not optimized. Photovoltaic production is lower.

How to wire up your solar power system

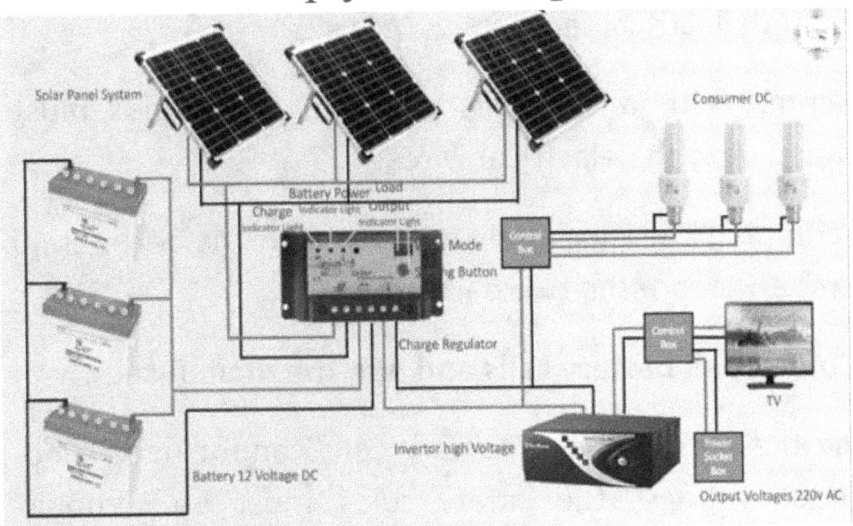

Like the panels and the batteries, the whole system can be wired in series and/or in parallel depending on the expected results. In series for a higher voltage (V), in parallel for a higher current (A), in series and parallel to increase both the voltage and the intensity of the system.

How to crimp

The crimping in electricity allows you to connect cable ends to an electric wire. An electrical wire must indeed be provided with a tip because of its composition: it is formed by a multitude of strands of copper. These tips ensure optimal connectivity by tightening the wire and all the strands that compose it, but also as an aspect of security.

The ends of the electrical wiring are fitted with an insulating collar which allows the insulation of the wire to be completed up to the electrical connection. This refers to pre-insulated electrical terminals.

To crimp with a crimping tool, the cable ends must be crimped onto the electrical wires

By crimping the electrical terminal with the wire, crimping pliers can secure the two parts.

1. Connect all battery cells and add the main fuse

Since the wiring of the battery cells cannot be carried out without voltage, the safety rules must be scrupulously observed.

Special care should be taken when handling the elements and acid.

Be reminded that it is imperative to use the safety equipment recommended in the installation instructions of the battery manufacturers (at least, clothing cotton and anti-static, safety glasses, appropriate handling gloves, insulated tools, eyewash), this list is not exhaustive.

Rule:

- If the batteries are connected in parallel the voltage remains the same but the capacity increases.

- If the batteries are connected in series, the voltage increases but the capacity remains the same.

- Only use batteries of the same brand, same type, and same age for these assemblies.

It is recommended to protect your batteries with a fuse. Install a fuse holder between the regulator and the battery, and an amperage fuse adapted to your panels. If they can send 50A in 12V (600W), then plan a 60A fuse.

2. Connect Solar Charge Controller To the Battery bank

The function of the charge regulator is to protect the battery against overcharging; it cuts off the current flow from the panel to the battery when the battery is fully charged.

If the charge controller has a USE or LOAD output, it is possible to connect users to this output. In this case, the charge controller will protect the battery from deep discharge by stopping use. However, it must be of the same voltage as the battery bank and must not have a current draw higher than the current which the regulator can withstand.

When it comes to the wiring between the charge controller and the batteries, there are two methods of wiring for the connection:

Basic method: The connection can be made with a battery regulator cable with blade fuse on 1 pole.

European Method: you may need to use a battery protection box conforming to UTE C 15-712-2 which is composed of gG fuses and a switch disconnector.

3. Connect individual panels to create a solar panel array

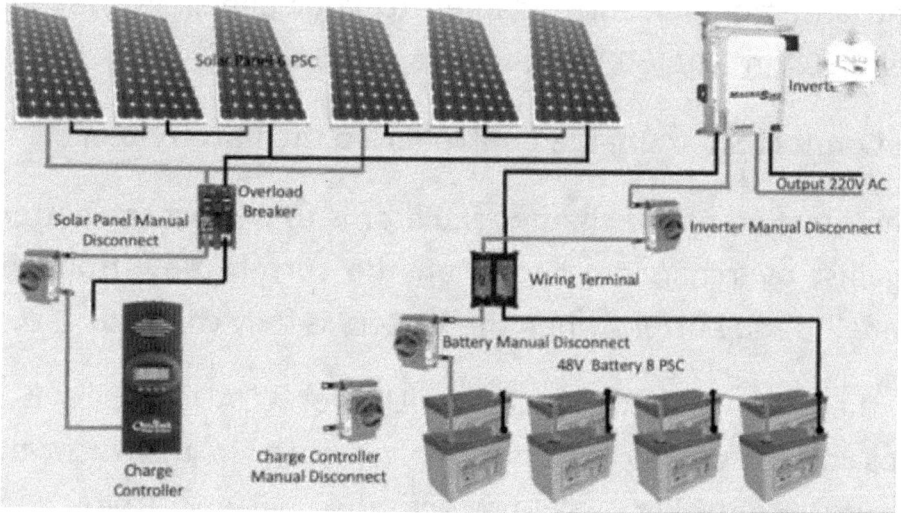

It is strongly advised to use solar panels of the same power for all your solar field.

Connection of solar modules in series:

When solar panels are connected in series, all the panels must have the same current. This connection makes it possible to add the voltages while the intensity in amperes is not modified. In this configuration, the (+) terminal of the solar panel is connected to the (-) terminal of the next panel, etc.

Connection of solar modules in parallel:

When solar panels are mounted in parallel, all the panels must have the same voltage. This connection makes it possible to add the intensities, while the voltage in Volts remains unchanged. In this configuration, the positive terminals (+) of all the solar panels are connected, as well as all the negative terminals (-). Parallel mounting requires the use of a junction box.

Connection of solar modules in series/parallel:

This is the connection you need when a certain power is required at the desired voltage. The rules for series connection and those for parallel connection apply in the case of series / parallel connection.

4. Pass Solar array wires through the roof and connect them to the Solar Charge Controller

The solar panels are installed on the roof, it remains to bring the cables inside to the charge controller. For that, we need a roof hatch.

To install your roof hatch

Clean the area

To better stick the roof hatch, the area must be cleaned. Use white spirit and a cloth to remove all dirt and degrease.

Pierce the roof

It is necessary to drill a hole in the roof which will allow you to pass the two cables, the positive and the negative.

Start by drilling a hole with the smallest metal drill, to get a drill of suitable size.

Then check that the two cables pass well. Do not make an excessively large hole because it can cause water infiltration.

Finally, clean the pieces and cut metal shavings well to prevent rust. Degrease the surface to be bonded again if necessary.

Paste

Put Sikaflex on the lower part of the roof hatch: type 291i, 292i, 512UV or 552 AT, depending on what you have in stock.

Position it and screw in with the supplied stainless steel self-drilling machines. Allow it to dry for about 24 to 48h, but do not forget to fix the upper part so that the water stays outside.

Pass the cables and close

Run the cables from the solar panels to your charge controller. Remember to tighten the screws at the bottom of the device to seal it. Finally, depending on the model put on the "hood" of the hatcher and screw it on.

At this point, your cables from the solar panels arrive at the charge controller. The charge controller must connect to the battery before connecting the solar panels to the charge controller.

As mentioned earlier, you can connect your panels in series or parallel. You may need an MC4 cable with T and Y connectors.

In parallel, your panels will need two "Y's" to connect the positive cables and the negative cables. Connect the "Y" terminal to the solar charge controller.

In series, you will need simple MC4 extensions because it will be necessary that a positive terminal of one panel connects to the negative terminal of the other. Then the two remaining cables (one + from one panel and one - from the other) will go to the charge controller.

5. Connect the inverter and Fuse block to the Battery bank.

Fuse Block installation

Firstly, install the Fuse block on the positive post of the Battery. Ensure you tighten up the terminal, but make sure it's not too tight.

Inverter installation

The first step is to remove the blusher on each terminal on the inverter. Place the red wire on the red terminal (positive) and screw. Repeat the same process for the black terminal (negative) as well. Remember to tighten up both terminals, but ensure they're not too tight. Place the caps on each terminal of the inverter for protection.

Connect the red wire from the inverter to the Fuse block located on the positive post on the Battery. For the negative post, slightly tap the end of the black wire on the negative post of the Battery to allow the inverter to charge up its internal capacitors. Connect the black end of the wire to the negative terminal of the Battery to complete the installing.

6. Battery Monitor Installation

A battery monitor set consists of a computer with a display and a shunt. The assembly comes down to placing the shunt in series in the battery line. On one side, it is connected to the battery terminal (positive or negative depending on the brand). The other side is connected to the mainline where the different producers arrive and the departure for consumers (electrical panel). On the shunt, two small screws are used to connect the measurement cable which is directly connected to the monitor. This assembly poses no difficulty. Once installed, it remains to configure it according to its battery bank.

Adding DC 12 Volt Appliances

A solar panel often supplies more voltage than is required. All dc appliances come with a maximum voltage of 12 volts. Some solar panels offer as much as 21 volts which are higher than that of a DC appliance. Despite this, you can still connect a DC 12 volt appliance directly to the panel and have it work. However, the appliance will only run for a short time.

To successfully use a DC appliance on your solar power system, you will need to use a voltage regulator. The voltage regulator works differently from the charge regulator which automatically charges the batteries of the panel till it's completely charged. The voltage regulator on the other hand simply converts high voltage to lower voltage to allow the system to run DC 12 volts appliances properly.

XT-60 Connector

The XT-60 connector helps to connect the wires of your solar panel to the MC4 cables of the system. The connector can carry up to 60 amps of power. However, the XT-60 connector does not use a mechanical connection as you are required to solder the solar panel wire to the connector.

Powering a Laptop Without an Inverter

Powering your laptop from a solar panel without the use of an inverter is very possible. Laptops require a voltage of about 19-20 volts to function properly. If you're using a DC to power the appliances in your van, you will be able to supply only as much as 12 volts. This, therefore, requires you to obtain an adapter that can help you to increase the voltage output to about 19 to 20 volts making it safe for you to power up your laptop. An example of a good adapter or converter is the PWR adapter which is available for purchase on Amazon. An inverter does almost the same thing as the adapter, but the adapter helps you to save power which is very essential if you're using a solar power system.

Adding efficient interior lights to your vehicle

The best and efficient light to add to the interior of your vehicle are LED lights. Conventional lights transform 95% of the energy used into heat and only 5% into light. It is the greater energy efficiency of LED lamps which allows substantial energy savings.

For equivalent lighting, LED lamps (the lighting mode used in traffic lights) consume 90% of energy less than conventional lights. Also, their lifespan is estimated between 50,000 and 100,000 hours (compared to 1,000 hours for a conventional bulb)

A LED bulb stays cold (or even slightly warm) whatever the duration of use. Finally, LED lighting offers a big advantage, compared to compact fluorescent bulbs, of not containing mercury, harmful to health and the environment.

Temperature regulation appliances

Many van owners are always in search of ways to run temperature regulation appliances like Air conditioners while using solar power systems. These appliances tend to consume a high amount of power, requiring you to pay much attention when using it with your solar power system.

Cooling your vehicle

The major cooling appliance used in vans is the air conditioner. This appliance is a high wattage appliance as it's compressor requires high current when running. The wattage of the cooling appliance is often based on its size. Also, the wattage of cooling appliances like the air conditioner varies when the appliance is running. This is because the compressor will automatically turn off when it reaches the temperature set by the thermostat. At the point where the compressor is turned off, electricity consumption is completely reduced

Heating Your Vehicle

Similarly to air conditioners, electric heater wattage has a requirement of up to 4000 watts. The high use of wattage is only needed until the van has reached the temperature set by the thermostat.

To be able to handle the high load demands, the inverters of the solar power system should be able to handle the high load. This means your inverter should be able to handle the power needs of up to 5000 watts.

Other methods

There are other free methods to regulate the temperature of your van

Install insulating shutters

Windows and windshields are a big waste of heat. To avoid this, we strongly advise you to invest or make insulating shutters for all your windows. We made them out of our multi-layered insulation scraps and saw a huge difference in installing them against the windows from inside the cabin. If you don't have a partition between the cabin and the passenger compartment, you can opt for an insulating curtain that will separate the two areas when you need them. This solution allowed us to maintain heat.

Gas heaters

Gas heaters are auxiliary heaters, not to be allowed to run continuously. They can be dangerous if basic precautions are not taken, namely:

Air out every day

Install a toxic gas detector

Pay attention to the risk of fire, not putting anything flammable near it

You can install a gas heater if you already have a good base of gas-powered equipment. The risk of carbon monoxide release is no higher than any other type of heating, and one of the major priorities is to be discreet.

How to use a Bulk Dc-Dc converter

The DC-DC converter converts a DC voltage like that of a battery (12V, 24V, or 48V) into a different DC voltage. There is also a DC-DC converter which delivers the same output voltage as input but stabilized and cleared of noise due to conversion thanks to galvanic isolation. The output voltage can be adjustable.

Choice and compatibility

To properly select a DC converter, you must know:

- The nominal input voltage of the battery

- The output voltage of the receiver

- The current or power of the receiver

1) Input voltage

The battery park voltage can be 12V, 24V, or 48V, and more rarely 36V or 96V. It is the DC voltage of the system. The larger the system, the higher this voltage to reduce the current. If you have a doubt, to know the voltage of your battery park, go to the page "installation of the battery".

2) Output voltage

The power supply of the receiver may require a DC converter if the voltage is different from that of the battery or if the power supply is said to be "sensitive". This means that it only tolerates small voltage variations and no interference such as electronic cards and telecommunications devices.

3) Nominal current

The converter will tolerate higher current over a short time to allow receivers to create a surge of current at start-up. This maximum current varies according to the models and the technology, it is, therefore, necessary to ensure that the peak at start-up can be supported by the converter.

Installation

The main sources of converter failure, apart from a compatibility or handling problem, are ambient temperature and dust. The energy conversion heats the device, the higher the charge, the more it heats. This excessive heat degrades the efficiency, and in extreme cases, puts the converter in fault. This is why the converters are fitted with a grille and a fan.

The charger must be installed in a dry and ventilated place. If dust collects on the air inlets, they should be cleaned.

Wiring

The choice of cable is essential for the proper functioning of the system. An insufficient cable section will cause a great loss of energy. The dimensioning of the cable is defined by the current and the voltage which crosses it as well as its length.

Check the polarity of the cables before each connection using a tester (multimeter) or by locating the cables with tape, for example. The common colors being red for (+) and blue for (-).

Safety instructions

Most often the flexible cable will be preferred (multi-strand thin), so make sure that all the strands of the cable are properly tucked into the connection terminal block. If strands are crushed, bent, or pulled out, repeat the connection.

Do not cut strands on a cable to reduce the cross-section. If the cable is too large for the terminal block, it should be replaced by a lower section. The maximum section tolerated by the regulator is indicated on the technical sheet.

The risks linked to strands protruding from the terminal block or to a bad connection are

Short circuit, fire, or burns.

Adding AC Appliances

AC appliances can not work directly with solar power systems. This is because solar panels produce power in DC which is unable to power AC appliances. However, you can always rely on an inverter to cover the DC power to AC.

Most of the appliances we use are powered by AC. Unlike DC, AC offers flexible power which can be transformed into different voltage outputs, thanks to the use of step up and step down transformers. It is due to this reason that many household appliances use AC power. Nonetheless, appliances that use DC power (batteries, cell phones, laptops, etc.) often come with adaptors containing built-in converters to convert AC power to DC.

Off-Grid Internet

Everyone is dependent on an internet connection for work or other reasons, and it can be important to be able to make a quick call. However, the van is far from everything, so you may not have a phone line and a very weak cell phone signal. However, a combined internet/telephone system requires less than 10 watts to maintain a constant high-quality connection. To improve reception, you can install a signal amplifier or a router with a high gain antenna.

4G LTE Router with High Gain Antenna

A 4G router is a great way to get access to internet connections when you are off Grid. Most 4G LTE routers come with a high gain antenna which you can mount outside your house. These antennas can receive signals when pointed in the direction of the nearest cell tower, about 8 km away.

From this antenna, a coaxial cable is connected to a signal amplifier (a box similar in size to that of a modem) installed in the van. The latter is connected to an indoor antenna that receives the cellular signal before communicating with the tower. Everything works very well, reducing a weak signal, almost non-existent, to a very strong signal (5 reception bars out of 5).

Smart Home Appliances

Smart home appliances are no longer new to us now. These connected devices have already swept the market, among which are connected sockets and connected appliances. Indeed, smart home appliances are a very flourishing trade, and consumers are more and more attracted by these new intelligent products.

Also, large brands like Samsung, Bosch, and LG are increasing the number of connected products. We thus find in their showroom refrigerators, dishwashers, multifunction ovens or even cooking robots that have the particular characteristic of being connected, being intelligent but also being durable and resistant as their brands promise. made.

Ease of installation

To install a connected electrical outlet, simply place it on your conventional outlet. This will allow you to connect the device connected to it. Connected sockets are thus an excellent way to switch off or on your lamps remotely or even to configure their switching on for a specific time slot. Same for your connected appliances or your radiator that you can turn on with one click.

Solar System Maintenance Schedule

Photovoltaic solar installations require very simple minimum maintenance. This maintenance is reduced to the following operations:

Panels: they require almost no maintenance, given their design: they have no moving parts and the cells and internal connections are trapped in several layers of protector. It is advisable to carry out a general inspection once or twice a year to ensure that the connections between the panels and the controller are in good condition and do not suffer from corrosion. In most cases, the action of rain avoids having to wash the panels; if necessary, simply clean them with water and a non-abrasive detergent.

Charge controller or controller: the simple operation of the controller significantly reduces maintenance. Damage to this type of equipment is relatively rare. The maintenance operations that can be carried out are as follows: visual observation of the state of the regulator; checking the connections and cables connected to the equipment; observation of the instantaneous values displayed via a voltmeter and an ammeter or directly on the LCD of the regulator for the most sophisticated models. These values provide clues to the behavior and state of the installation.

Battery: this is the element of the installation that requires the greatest attention; its lifespan will depend directly on how it is used and how well it is maintained. The usual operations which must be carried out on a conventional battery (certain types of batteries, such as AGM batteries, for example, require no maintenance) are the following:

-Checking the electrolyte level (approximately every 6 months): it must be kept within the margin between the "Maximum" and "Minimum" marks. If there are no such marks, the correct electrolyte level is 20 mm above the separator protector. If a lower level is observed in some of the elements, they must be filled with distilled or demineralized water. They should never be filled with sulfuric acid.

- When carrying out the previous operation it is important to check the condition of the battery terminals; the terminals must be cleaned (removal of traces of sulfate) and it is advisable to cover, with neutral vaseline, all connections.

- Measurement of the density of the electrolyte (if a hydrometer is available): with the fully charged battery, it must be 1.240 +/- 0.01 at 20 degrees Celsius. Densities must be similar in all elements. Significant differences in an element can reveal the damage.

Odds and Ends

Efficient Computer Options

Even though it consumes much less than a fridge, your computer consumes energy. Admittedly, this consumption is reduced compared to certain household appliances. However, the most powerful appliances, especially the most recent, cannot boast of their energy efficiency. If we add to this a bad technical configuration, that's enough to consume a lot of energy.

Here are some quick and easy tips to avoid overconsumption of electricity due to your computer.

- **Turn off the screen, and avoid the "savers"**

A screen saver will give your computer a personal touch, but it's an unnecessary waste of energy. Neither the screensaver nor leaving the screen on will help you save energy. So turn off the screen as much as possible to extend its lifespan and, of course, save energy.

- **Standby is recommended**

Whatever the operating system, your computer has sleep and hibernation options, a way to dramatically reduce idle power consumption. It does this while keeping your PC ready to get things done as soon as you are sitting in front of the screen. When you are not going to use it, even if it is only a few minutes, it is advisable to use these options, including programming its automatic activation.

- **Your hard drive consumes a lot of energy**

This is not something that many users are considering and yet it is one of the most important factors. Computer memory, that is, the hard drive, uses mechanical systems that require more energy than an SSD. Changing the component in favor of a solid-state drive (SSD) may require a small expense, but you will gain speed and energy efficiency.

- **Beware of peripherals**

Almost all of the devices you connect to the computer, such as a mouse or external hard drive, use the computer itself as a power source.

It is not a question of buying components that connect directly to the outlet, but of maintaining this type of power and taking advantage of more efficient components.

- **Other ways to save energy with your computer**

In the Power Options of your computer, you will find the performance settings to adjust the extent to which the computer is allowed to provide maximum power. If you use such a configuration, your computer will still be ready to run the heaviest programs, but with higher performance than you need most of the time and consuming unnecessary resources. Of course, keeping the volume and brightness low will help you, on a desktop and laptop computer, to reduce your power consumption.

Phantom Loads

Up to 10% of your electricity corresponds to the consumption of electrical devices that are turned off (at least in appearance). It is invisible energy consumption, which has several names: ghost charges, vampire consumption, standby mode (or standby mode), loss of electricity, or leakage of electricity.

In this century, "off" does not necessarily mean "off". Even when switched off, several household appliances and electronic devices continue to consume electricity to power various functions (e.g. clock, indicator lights, touch buttons, and reception of network or remote control signals).

How to find Phantom Loads

Eliminate almost all of your phantom loads by turning off all your devices when you're not using them. It's simple, and it's the only way to make sure your electronic devices don't consume electricity. Unplug them from the wall outlet or plug them into a power strip that can be turned off.

- Phantom loads represent 5-10% of your electricity costs.
- Most appliances are usually used between 3 and 30 minutes a day; the rest of the time, they are in standby mode.
- Electronic devices are in standby mode about 75% of the time or about 6,500 hours per year.
- Up to 40% of the annual consumption of all electronic devices occurs when they are switched off.
- Sometimes, some devices that are used very little consume more electricity in a year when they are turned off than when they are turned on.
- A home contains on average 20 to 40 devices generating ghost charges.
- The electricity consumption of these devices can reach almost 30 W but also be limited to 0.5 W.

Storing a Solar Power System

When your solar power system is not in use, you must store it properly to avoid damage to the system. However, storing a solar power system does not require much work or skills. Firstly, you have to ensure that all wires are properly disconnected. Ensure that your battery is charged well above 50% before storing it for a long time. This will make sure that the battery does not completely die out when placed in storage for long periods.

Also, remember to store the batteries, inverter, Charge Controller, and other equipment in a cool and dry place, away from humidity. As for the panels, you can decide to cover it up or leave it open depending on your preference. However, if you're concerned about extreme weather, you can cover up the solar panel or simply take it down.

Connecting different types of solar panels together

In summary, connecting panels of different brands and power are possible provided you have the right voltages. However, if your panels are mounted in series or parallel, you must take into account the layout and dimensions of your panels. As a reminder, the layout is the art of choosing the orientation and the number of panels to be installed on a given surface.

Be careful, despite the brand difference not being important, the technology of the panel should not be mixed with another technology. It is not recommended to install a panel with monocrystalline cells on the same string as a panel with polycrystalline cells. Do not install a polycrystalline and monocrystalline panel on the same solar system.

Connecting different Solar Charge controllers to one battery bank

You can always connect different charge controllers to a single battery bank if a single controller is unable to handle the charge from the solar panels. The voltages from the two Charge controllers can be different and may need to be connected in series.

Also, the voltage of your battery bank has to be lower than the voltage of the Charge Controller array. This will allow the Battery to be charged conveniently by the solar power system. The dc breaker box is where you can combine all the Charge controllers before connecting it to the battery bank.

Solar Electric Cooking and food preparation

Cooking with a solar power system is also possible and may require a lot of energy consumption. However, there are few factors or solutions to help reduce the high energy consumption for cooking.

Induction hobs consume 40% less energy than conventional electric hobs. Their higher purchase cost is offset by the difference in energy consumption.

- Thaw your food in a refrigerator before reheating it. So your fridge takes advantage of the coolness and the power consumption of the microwave is reduced.

- Space out the oven pyrolysis: pyrolysis consumes a lot of energy. It is best to do this only once a year, cleaning the oven manually for the rest of the year.

- Switch off the plates a little before the end of cooking: this is a recommendation for cooking on electric plates. Indeed, they continue to heat approximately 15 minutes after their extinction. You can also do the same with the oven. Overconsumption is estimated at 15%.

- Avoid opening the oven during cooking: several degrees are lost each time the oven is opened. It is therefore preferable to switch on the interior lighting.

- Cook food with a lid. It keeps the heat inside the pan and therefore reduces the cooking time. In addition to the energy-saving, this saves time since the food is cooked faster.

- Descale your kettle: the tartar increases the healing time of the water (this can reach 50% additional time) and damages the appliances. tip: use vinegar to descale the electric kettle or the iron, or diluted phosphoric acid for washing machines.

Solar Water Heating

Solar water heating is a complement to electric water heating and other gas water heating. It allows the occupants of a dwelling to cover a large part of their consumption of domestic hot water, thanks to solar energy. To do this, it uses the thermal conversion of solar radiation.

The solar water heater can also be used to heat your van. Combined with a radiant heating system with water or even forced air, it is very effective in reducing energy consumption and provides healthy and gentle warmth to the inhabitants.

It is important to differentiate between active solar technologies and the use of passive solar energy. Active solar works thanks to mechanical devices and devices that recover and store heat. Passive solar is a design technique that favors the exposure of the house to the rays of the sun while adequately protecting the faces exposed to cold winds and shade. The solar water heater is therefore considered to be an active solar system.

Contrary to popular belief, installing a solar water heater is not a child's play. Also, unless you have solid skills in electricity, plumbing, and carpentry, it would be better to call a qualified professional with specific experience in installing the solar water heater.

The sensor is preferably placed full south at an optimum annual inclination of 40 to 50 degrees on the horizontal. Tolerance for architectural reasons: 25 to 30 degrees. If the slope of your roof is not facing due south, you can fix the sensor on the wall or the ground using a specially designed chassis. You must however ensure that none of the elements in the surrounding will shade the sensor during the day.

Should I install a Battery Isolator?

Yes, a battery isolator is a nice device to install in your camper van. A battery isolator is an on / off switch that disconnects the starter battery in a vehicle. Another name for a battery isolator is the "cut-off" battery switch. This isolator can prevent battery drains and also acts as an anti-theft device. It will allow you to activate a key and disconnect the battery from the starter system. Without a key to the isolator, the battery cannot be reconnected and the car will not run.

Increasing Solar output by reflecting light onto your solar panel

Reflecting light onto your solar panel increases the efficiency of solar panels. This is because photovoltaic cells benefit from the reflection of the sun's rays. In practice, this results in better electrical efficiency compared to little or low reflection. Having a reflective roof is also a great way to reflect light onto our solar panel.

A reflective roof covering increases the performance of solar panels in two ways. On the one hand, the coating remains cooler, even when the sun is shining brightly. You should know that the temperature can rise to 80 degrees in summer on a black roof. With a white roof, your solar installation remains much cooler. And this has a positive impact, because the hotter the silicone in solar cells, the lower their efficiency - count an efficiency loss of around 1% per 2-degree increase.

Also, solar cells benefit from the reflection of the sun on the roof surface. A small part of the solar energy produced by the panels comes from the sun's radiation through the roof. On a sunny day, therefore, electricity production increases.

About the Author

Larry Barone has over 10 years in setting up renewable energies systems and technological research for clean and more efficient energy. He works so hard to simplify the installation and maintenance of complex solar power systems, making it very easy for hobbyist to install them at homes, vans, RVS, boats etc.

He lives in New York, USA. He is happily married with two kids.

www.ingramcontent.com/pod-product-compliance
Lightning Source LLC
Chambersburg PA
CBHW070259220526
45465CB00004B/1661